INTERNATIONAL TD CRAWLERS 1933 THROUGH 1962

PHOTO ARCHIVE

IN MEMORIAM

PETER LETOURNEAU
1950 - 1997

We dedicate this book as a memorial to our friend, partner, and fellow worker Peter Letourneau, who died driving a 1926 3 litre Bentley, a passion close to his heart. We will miss him and wish him God Speed.

Shawn Glidden, Theresa Piemonte, Rick and Debbie Seymour, Tom Warth

INTERNATIONAL TD CRAWLERS 1933 THROUGH 1962

PHOTO ARCHIVE

Photographs from the
McCormick-International Harvester Company Collection

Edited with introduction by
P.A. Letourneau

Iconografix
Photo Archive Series

Iconografix
PO Box 446
Hudson, Wisconsin 54016 USA

Books in the Iconografix *Photo Archive Series* are offered at a discount when sold in quantity for promotional use. Businesses or organizations seeking details should write to the Marketing Department, Iconografix, at the above address.

Library of Congress Card Number 97-70640

ISBN 1-882256-72-7

97 98 99 00 01 02 03 5 4 3 2 1

Printed in the United States of America

PREFACE

The histories of machines and mechanical gadgets are contained in the books, journals, correspondence, and personal papers stored in libraries and archives throughout the world. Written in tens of languages, covering thousands of subjects, the stories are recorded in millions of words.

Words are powerful. Yet, the impact of a single image, a photograph or an illustration, often relates more than dozens of pages of text. Fortunately, many of the libraries and archives that house the words also preserve the images.

In the *Photo Archive Series,* Iconografix reproduces photographs and illustrations selected from public and private collections. The images are chosen to tell a story—to capture the character of their subject. Reproduced as found, they are accompanied by the captions made available by the archive.

The Iconografix *Photo Archive Series* is dedicated to young and old alike, the enthusiast, the collector and anyone who, like us, is fascinated by "things" mechanical.

Iconografix continuously seeks collections of archival photographs for reproduction in future books. We require a minimum of 120 photographs per subject. We prefer subjects narrow in focus, i.e., a specific model, railroad, racing venue, etc. Photographs must be of high-quality, suited to reproduction in an 8x10-inch format. We willingly pay for the use of photographs.

If you own or know of such a collection, please contact: The Publisher, Iconografix, PO Box 446, Hudson, Wisconsin 54016.

The photographs and illustrations that appear in this book were made available by the State Historical Society of Wisconsin. The Society is the official repository for records of the International Harvester Company and its nineteenth-century predecessor, the McCormick Harvesting Machine Company.

The McCormick-International Harvester Company Collection contains nearly 4,000 cubic feet of family papers and business records, including technical publications, advertising literature, engineering and promotional photographs, posters, and films. Its cataloguing, preservation, and administration are funded through an endowment established in 1991 by Brooks McCormick.

Use of the collection is not generally restricted. However, due to its size and complexity, interested persons are encouraged to contact the Society in advance, at 816 State Street, Madison, Wisconsin 53706.

TD-18 with Heil scraper at work on an Alabama highway project, 1947.

INTRODUCTION

The earliest International Harvester crawler tractors were modified wheel tractors equipped with full or half tracks manufactured by allied suppliers such as Trackson, Hadfield-Penfield, Moon Track, and Mandt-Freil. By 1928, IH was building its own tracks, and in 1930 introduced its first purpose-built crawler tractor, the TracTracTor Model 15 or T-15. In 1931, the T-15 was supplanted by the T-20. A year later, IH introduced the TracTracTor TA-40.

Prior to 1933, all TracTracTors were powered by spark-ignition engines. In 1933, IH introduced the TD-40, a diesel-powered version of the T-40 (the TA-40's successor). The engine, a 4¾ x 6½-inch 4-cylinder, was equipped with magneto, carburetor and spark plugs, as well as fuel injection pump and injectors. The engine was started on gasoline and, once up to speed, operated on diesel fuel. The TD-40, equipped with 5-speed transmission, developed a maximum 53.46 brake and 48.53 drawbar horsepower, and achieved maximum drawbar pull of 10,487 pounds in its Nebraska Tractor Test. The TD-40 remained in production until 1939.

The TD-35 was introduced in 1937 and was built until 1939. It featured a 4½ x 6½-inch 4-cylinder diesel that was started on gasoline and switched over to diesel fuel in the same manner as the TD-40. The TD-35 was also equipped with a 5-speed transmission.

In 1939, a new series of diesel crawlers was introduced. Styled in a streamlined fashion by the famed industrial designer Raymond Loewy, the four tractors featured new engines, transmissions, and main frames, as well as numerous mechanical, electrical, and hydraulic improvements. The TD-6, smallest of the new line, was built from 1939 through 1969. It was equipped with a 3⅞ x 5¼-inch 4-cylinder engine and 5-speed transmission. The TD-6 developed maximum 34.54 brake horsepower and achieved maximum drawbar pull of 7,160 pounds in its Nebraska Test. In 1956, the TD-6/61 Series was introduced. The 61 Series' 4 x 5¼-inch 4-cylinder engine was rated at a maximum 48.99 belt horsepower, and was mated to a 5-speed transmission. The 61 Series was succeeded by the 62 Series in 1959. The principal differences in this later series were the variety of transmissions and tractor configurations offered. The new models included: TD-62 Manual Shift; TD-62A Manual Shift Agricultural; TD-62S Manual Shift Sideboom; TD-62L Manual Shift Loader. Production of 62 Series crawlers ended in 1969.

The TD-9, equipped with a 4⁴⁄₁₀ x 5½-inch 4-cylinder diesel and 5-speed transmission, developed maximum 43.93 brake horsepower and achieved maximum drawbar pull of 9,014 pounds in its Nebraska Tractor Test. The TD-9/91 Series was introduced in 1956 and was built through 1959. It was succeeded by the 92 Series, which was built through 1962.

The TD-14 was built from 1939 through 1949. It was equipped with a 4¾ x 6½-inch 4-cylinder engine and 6-speed transmission. The TD-14 developed maximum 61.56 brake horsepower and achieved maximum drawbar pull of 13,426 pounds in its Nebraska Tractor Test.

The TD-14's successor, the TD-14A, remained in production until 1955. The TD-14A offered a boost in power of approximately 10 horsepower. The TD-14A was succeeded by the TD-14A/141 Series, built in 1955 and 1956, and the TD-14A/142 Series, built from 1956 to 1958.

The TD-18 was built from 1939 through 1948. It featured a 6-speed transmission and 4¾ x 6½-inch 6-cylinder engine. The TD-18 developed a maximum 80.32 brake and 72.38 drawbar horsepower and achieved maximum drawbar pull of 18,973 pounds in its Nebraska Tractor Test. The TD-18's successor, the TD-18A, remained in production until 1955. The TD-18A was succeeded by the TD-18A/181 Series, built in 1955 and 1956, and the TD-18A/182 Series, built from 1956 to 1958.

TD-24 production began in 1947. Equipped with an 8-speed transmission and 5¾ x 7-inch 6-cylinder engine, it developed a maximum 138.13 drawbar horsepower and achieved a maximum drawbar pull of 33,714 pounds. The TD-24 weighed nearly two tons more than a Caterpillar D-8. Its maximum drawbar pull exceeded that of the D-8 by more than 5,000 pounds. In 1955, IH introduced the TD-24/241 Series Gear Driven tractor and the TD-24/241 Series Torque Converter tractor. Both of the 241 Series tractors were built through 1959.

Although International Harvester continued to build crawler tractors beyond 1962, the McCormick-International Harvester Company Collection includes few photographs of the machines built after the early 1960s. Consequently, we are only able to offer limited coverage of the models which follow, all introduced prior to 1963. The TD-15 was introduced in 1958 and was built through 1962, when it was replaced by TD-15 Series B. The TD-15 was equipped with a 4⅝ x 5½-inch 6-cylinder engine and 6-speed transmission. The TD-20 200 Series was introduced in 1958 and built through 1960. It was succeeded by the turbocharged 201 Series, built in 1961 and 1962. Equipped with a 4¾ x 6½-inch 6-cylinder and 6-speed transmission, the 201 Series produced a maximum 110.48 horsepower and maximum drawbar pull of 28,236 pounds in its Nebraska Tractor Test. The TD-340 was equipped with the same 3 11/16 x 3⅞-inch 4-cylinder engine fitted to the Farmall 340 Diesel, and with a 5-speed transmission. A small track machine, it produced 32.83 maximum drawbar horsepower in its Nebraska Tractor Test. The TD-340 was introduced in 1959 and was discontinued in 1965. The TD-25/250 Series was built from 1959 to 1962 in both gear driven (8-speed manual shift) and 4-speed torque converter models. A 4-speed power shift unit and a 4-speed manual shift with torque converter unit were introduced to the TD-25 Series B in late 1962. With 8-speed transmission and turbocharged 5⅜ x 6-inch 6-cylinder engine, the TD-25 produced maximum drawbar pull of 47,244 pounds at 184.68 horsepower in its Nebraska Tractor Test. The Series B 4-speed Power Shift tractor was rated by IH at 230 horsepower. The TD-30, the last and biggest of the TD tractors featured in this book, was introduced in 1962. It was offered as a 320-horsepower 4-speed Power Shift model or 280-horsepower 8-speed manual shift model. The TD-30 was never tested at Nebraska.

Crawler tractors are the subject of several of the most popular books in the *Photo Archive Series*, including our best-seller *International TracTracTor Photo Archive.* This sequel seemed a natural, as the McCormick-International Harvester Company Collection includes so many great images of TD series tractors. We hope that you will find the photographs as fascinating as we find them.

Illustration of the TracTracTor Diesel Model 40 (TD-40), 1933.

Two views of a TD-40 with Adams scraper grading an unsurveyed county road, Hillsdale County, Michigan, 1938.

TD-40 with Bucyrus-Erie bulldozer skidding logs, 1937.

TD-40 used to haul and level garbage in a Chicago, Illinois municipal dump site.

TD-40 with a Bucyrus-Erie bulldozer in an Illinois coal mining operation, 1937.

Widetread TD-40 pulling three 7,000 lb. rollers over newly broken land on a 100,000 acre Manitoba farm, 1937.

TD-40 and Bucyrus-Erie two-wheel scraper building a road in Texas, 1938.

TD-40 and McCormick-Deering no. 51 Hillside Combine operating near Weston, Washington, 1937.

TD-40 with a Bucyrus-Erie bulldozer on a road project near Baysville, Ontario.

TD-40 with a Bucyrus-Erie bulldozer on a road project near Baysville, Ontario.

TD-40 on a terracing operation for Limestone County, Alabama, 1937.

TD-40 and two-wheel Bucyrus-Erie scraper on a road construction project near West Lancaster, Michigan, 1938.

TD-40 with a Bucyrus-Erie bulldozer building an artificial lake on Timber Trails Golf Course, Hinsdale, Illinois.

TD-40 pulling stumps. Hammon, Louisiana, 1938.

Illustrations of the TracTracTor Diesel Model 35 (TD-35), 1937.

INTERNATIONAL HARVESTER
TRACTRACTOR
INTERNATIONAL HARVESTER
TRACTRACTOR
DIESEL

Shifting a TD-35 engine to “gas” before starting.

Crank-starting a TD-35.

TD-35 widetread.

Belt pulley attachment on a TD-35.

TD-35 with orchard seat and fenders pulling a 9¾-foot covered dish harrow in a California orange grove.

TD-35 hauling logs near Winnipeg, Manitoba.

Two views of a TD-35 cultivating in an orchard belonging to T.G. Bright Wine Company, St. Catherines, Ontario.

TracTracTor
35
DIES

TD-35 pulling a 2-row grape sprayer in a vineyard belonging to T.G. Bright Wine Company, St. Catherines, Ontario.

TD-35 cleaning out a streambed near Vancouver, British Columbia.

TD-35 with cab and snow removal blade.

TD-35 pulling a 49-marker grain drill.

TD-35 pulling a 2-row beet puller on an 850-acre Clarksburg, California farm.

TD-14, circa 1947.

A TD-14 being fueled. This Minnesota farmer liked his TD-14 “because he (could) operate on black gumbo a few days sooner… (which helped) win the race with the weather.”

TD-14 and 15-foot Pacific Levaplane leveling ground prior to planting barley near Five Points, California.

TD-14 with Bucyrus-Erie bulldozer in a coal mining operation.

TD-14 with Bucyrus-Erie dozer and Bucyrus-Erie G-58 scraper on a dam project at IH's Hinsdale Farm, 1948.

TD-14 with Superior pipe boom repairing blast damage near Texas City, Texas, 1947.

TD-14 with Isaacson bulldozer leveling site around a newly constructed building in Valparaiso, Indiana.

TD-14 skidding logs in Routt National Forest near Walden, Colorado, 1946.

TD-14 with Bucyrus-Erie bulldozer and Carco winch clearing for a road northwest of Fort Williams, Ontario.

TD-14 pulling a wagon loaded with cotton.

TD-14 with Hough loader dumping rock into a crusher.

TD-14 with Luther oil field winch moving a Parkersburg cable tool drill in a Pennsylvania oil field, 1941.

TD-14 with sheepsfoot tamper, and an Adams motorgrader powered by an IH UD-14 engine, 1948.

TD-14 with Isaacson *Trackdozer* and WO-14 winch skidding logs in an Oregon forest. 1941.

TD-14 with LeTourneau scraper building a sump pit in preparation for drilling a new oil well for Western Gulf. In the background a conductor pipe is being set, 1948.

TD-14 with Bucyrus-Erie bulldozer building road for a pipeline operation, 1941.

A 1953 TD-14A, successor to the TD-14. The TD-14A was built from 1949 to 1958.

TD-65 (original designation for the TD-18) and Euclid wagon working on Sky Line Drive, Charlottesville, Virginia.

DIESEL
DIESEL

TD-65 and Adams grader on a road construction project, 1939.

TD-18 and Bucyrus-Erie scraper on a New York state road construction project, 1939.

Three TD-18s, one pulling a Bucyrus-Erie S-90 scraper, in a coal stripping operation for Corona Coal, Hepzibah, West Virginia, 1945.

TD-18 with Superior pipe boom pulling a Caterpillar D4 out of the mud along a canal bank.

TD-18 transporting 90 tons of freight for a Canadian contractor, 1941.

TD-18 with Bucyrus-Erie bulldozer placing a pump in the sump pit of a coal operation.

TD-18 with seven-yard wagon at work in the Belvedere Mountain open pit asbestos mine, Eden, Vermont, 1941.

TD-18 dumping a load into a crusher at the Belvedere Mountain open pit asbestos mine near Eden, Vermont, 1941.

TD-18 with Heil angle dozer working on a riverbed relocation project.

TD-18 with 32-inch Cyclone weeder in 300-acre bean field. The weeder cut subsoil roots of morning glories and pulverized the soil. Oceanside, California, 1949.

TD-18 with Superior pipe boom lays 30-inch pipe for a natural gas pipeline.

TD-18 with Hughes-Keenan Roustabout crane.

TD-18 with Carco arch and Isaacson winch in forest west of northern California's Lassen Park, 1946.

TD-18 with Isaacson arch in a Douglas Fir logging operation.

Experimental Isaacson wheel tractor based on a TD-18, 1946.

TD-18A (181 Series) with Allen cab. The TD-18A (181 Series) was built in 1955 and 1956.

A winterized TD-18 (182 Series) with enclosed cab, as built for the US Army Corps of Engineers. The TD-18 (182 Series) was built from 1956 to 1958.

TD-18 (182 Series) with IH 18D-2 bulldozer.

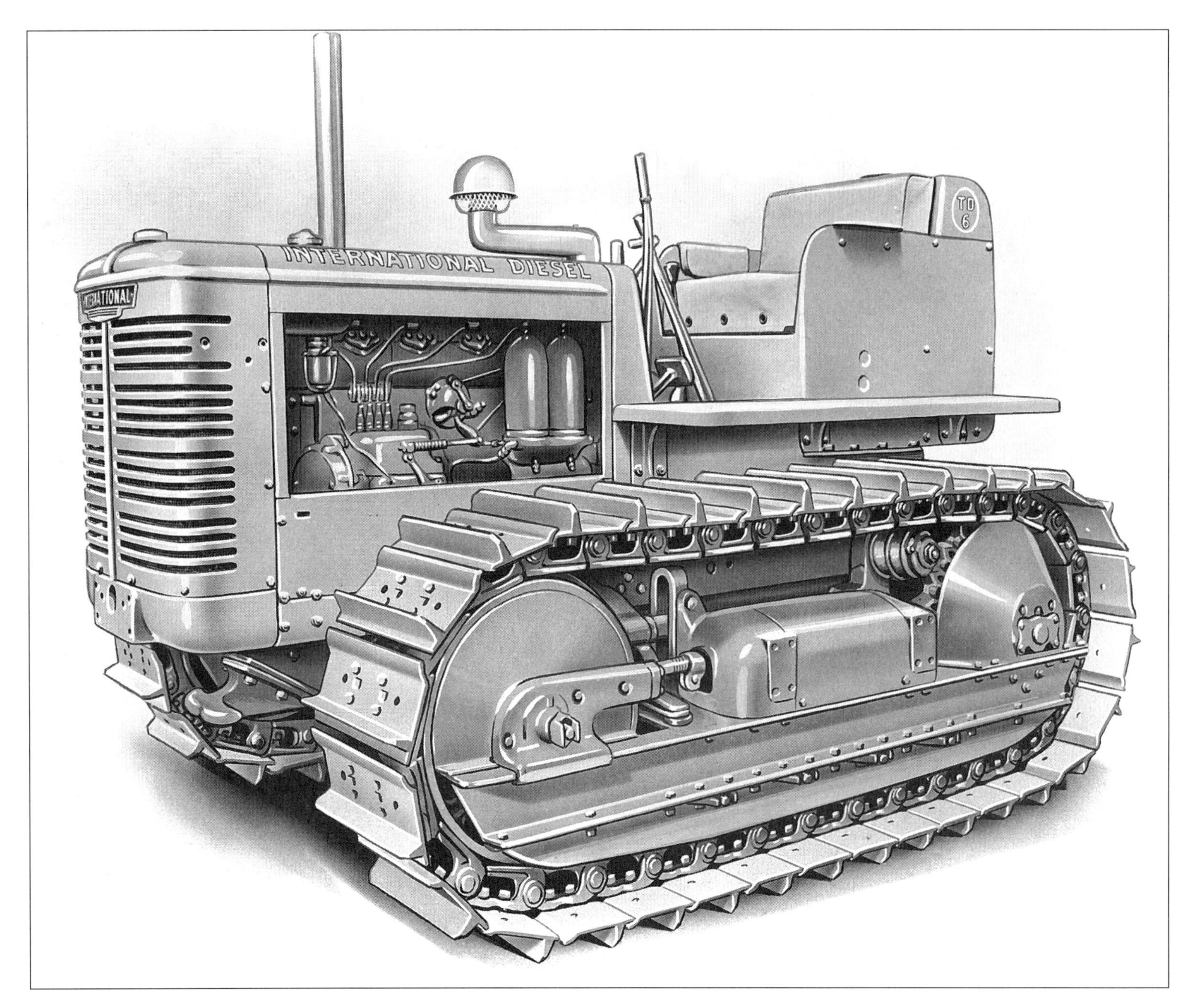

TD-6, circa 1947.

TD-6 with Heil hydraulic angle dozer backfilling a gas line ditch on a Milwaukee, Wisconsin street.

TD-6 with Hough ½-yard loader clears dirt from the shaft opening of a lead ore mine, Joplin, Missouri, 1944.

TD-6 with Bucyrus-Erie dozer shovel at work in a clay field, 1945.

Front and rear views of a TD-6 orchard tractor, 1952.

TD-6 with Killefer rotary scraper.

Diesel-powered International TD-6 building fire guards near Goleta, California, 1941.

TD-6 hauling logs at an Indiana mill, 1946.

TD-6 with Bucyrus-Erie dozer shovel excavating a basement for a home, Sheboygan, Wisconsin, 1947.

TD-6 at work on a 13-mile pipeline maintenance project. The side boom held the pipeline, while a crew at the front of the machine cleaned it, and a crew at the rear of the machine applied a protective paint coat, 1941.

TD-6 in a logging operation near Crosby, Mississippi, 1940.

TD-6 (61 Series) with IH 6D4 bulldozer. The TD-6 (61 Series) was built from 1956 to 1959.

TD-9, circa 1947.

TD-9 with Hough bulldozer shovel excavating a basement for a home, Omaha, Nebraska, 1947.

TD-9 with Trackson pipe boom on a pipeline reclaiming project, 1947.

TD-9 with Bucyrus-Erie dozer backfilling on a sanitary sewer project in Wichita, Kansas, 1947.

TD-9 with Hughes-Keenan fork lift, July 1949.

TD-9 with Drott skid loader loading pulpwood logs, 1947.

TD-9 with a Super Crane installing a new carrier on a slack pile at the Hudson Coal Co., Scranton, Pennsylvania.

TD-9 with Isaacson winch moving large rock from an Arizona pipeline right-of-way, 1947.

TD-9 with a Southwest loader digging clay from a bank and loading it into a dump truck.

TD-9 and all-metal Evans & Busch trailer loaded with pulpwood, 1947.

TD-9 with ½-yard front loader working in a crushed rock operation, 1944.

TD-9 with Isaacson winch towing a compressor along an Arizona pipeline right-of-way, 1947.

TD-9 with Bucyrus-Erie dozer shovel loading clay into a grinder at a brick and tile plant.

TD-9 (91 Series) with IH 964 bulldozer. The TD-9 (91 Series) was built from 1956 to 1959.

A TD-24, circa 1947.

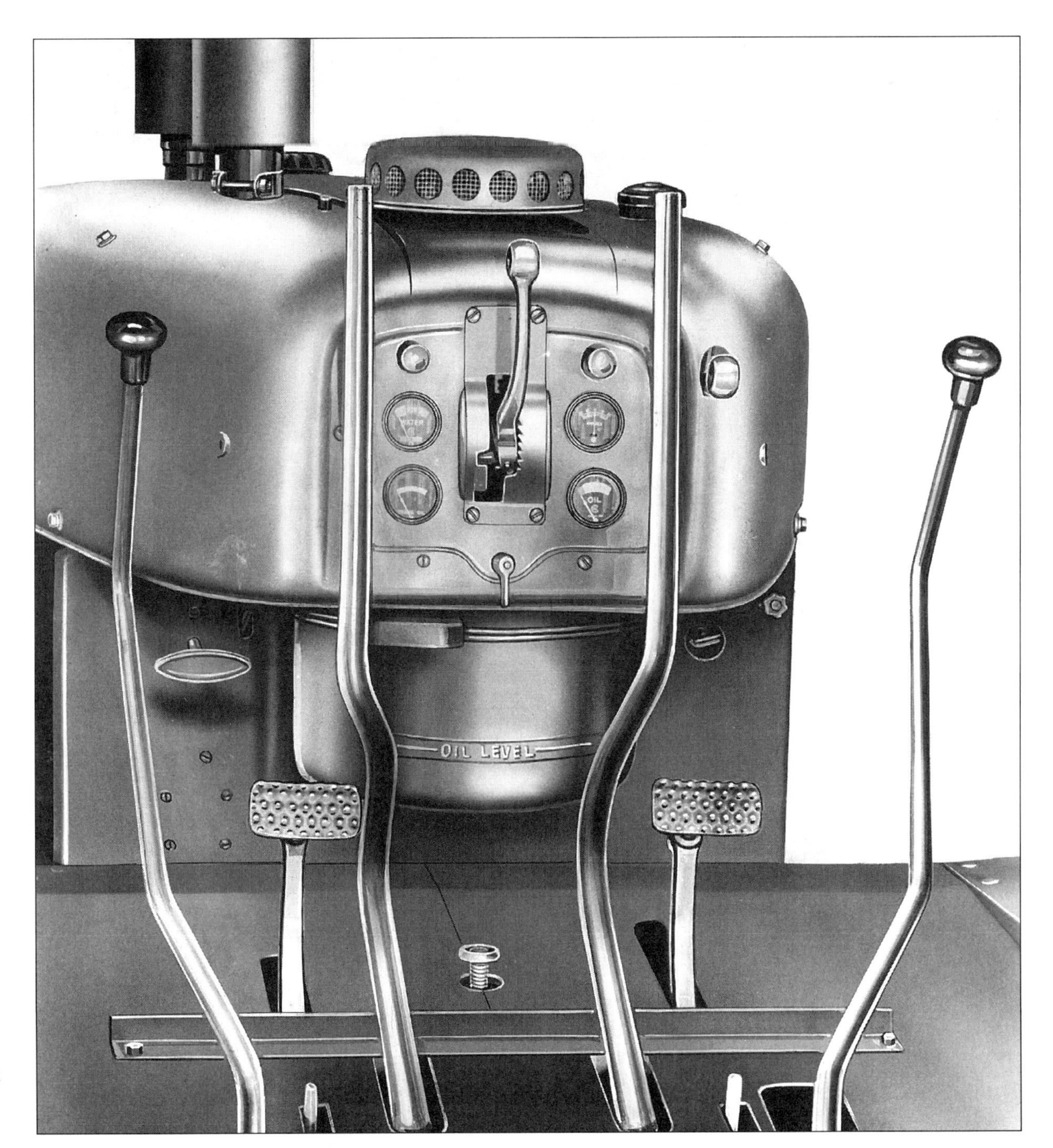

The operator controls on a TD-24, circa 1947.

TD-24 with Superior side boom on a California pipeline operation, 1949.

Skidding logs with a TD-24.

TD-24 with Bucyrus-Erie bulldozer on construction of an access road to a New York damsite, 1948.

A TD-24 delivered to the United States Navy in 1952.

A winterized TD-24 with enclosed cab delivered to the United States Navy in November 1952.

TD-24 with Bucyrus-Erie bulldozer making a walkway for a P&H Model 1055 dragline.

The first TD-24 sold overseas on a construction site in Saudi Arabia.

TD-24 with Superior pipe boom laying 20-inch pipe through the Texas desert.

Two TD-24s working on the Republican River Medicine Creek Dam, north of Cambridge, Nebraska.

A TD-24 sits on a canal bank, as a crew of men probe the waters for a newly lain pipeline, 1948.

TD-15 with hydraulic tilt dozer. The TD-15 was built from 1958 to 1962.

TD-20 with cable tilt dozer. TD-20 200 and 201 Series were built from 1958 to 1960, and 1961 to 1962, respectively.

TD-340, introduced in 1959.

A 1960 TD-25 forestry tractor. The TD-25 was built from 1959 to 1962 in the Gear-Driven 250 Series. The 250 Torque Converter Series was also offered the same years.

A 1961 TD-25BP (Series B Power Shift) with 4101-7 dozer with hydraulic tilt. The TD-25 Series B was offered with manual shift, manual shift with torque converter, and power shift.

TD-30 Torque Converter with Hydraulic Tilt 30 Dozer and Greenville Ripper. The TD-30 was introduced in 1962.

TD-30 Power Shift dozer, introduced in 1962.

The Iconografix Photo Archive Series includes:

AMERICAN CULTURE

AMERICAN SERVICE STATIONS 1935-1943	ISBN 1-882256-27-1
COCA-COLA: A HISTORY IN PHOTOGRAPHS 1930-1969	ISBN 1-882256-46-8
COCA-COLA: ITS VEHICLES IN PHOTOGRAPHS 1930-1969	ISBN 1-882256-47-6
PHILLIPS 66 1945-1954	ISBN 1-882256-42-5

AUTOMOTIVE

FERRARI PININFARINA 1952-1996	ISBN 1-882256-65-4
GT40	ISBN 1-882256-64-6
IMPERIAL 1955-1963	ISBN 1-882256-22-0
IMPERIAL 1964-1968	ISBN 1-882256-23-9
LE MANS 1950: THE BRIGGS CUNNINGHAM CAMPAIGN	ISBN 1-882256-21-2
LINCOLN MOTOR CARS 1920-1942	ISBN 1-882256-57-3
LINCOLN MOTOR CARS 1946-1960	ISBN 1-882256-58-1
MG 1945-1964	ISBN 1-882256-52-2
MG 1965-1980	ISBN 1-882256-53-0
PACKARD MOTOR CARS 1935-1942	ISBN 1-882256-44-1
PACKARD MOTOR CARS 1946-1958	ISBN 1-882256-45-X
SEBRING 12-HOUR RACE 1970	ISBN 1-882256-20-4
STUDEBAKER 1933-1942	ISBN 1-882256-24-7
STUDEBAKER 1946-1958	ISBN 1-882256-25-5
VANDERBILT CUP RACE 1936 & 1937	ISBN 1-882256-66-2

TRACTORS AND CONSTRUCTION EQUIPMENT

CASE TRACTORS 1912-1959	ISBN 1-882256-32-8
CATERPILLAR MILITARY TRACTORS VOLUME 1	ISBN 1-882256-16-6
CATERPILLAR MILITARY TRACTORS VOLUME 2	ISBN 1-882256-17-4
CATERPILLAR SIXTY	ISBN 1-882256-05-0
CLETRAC AND OLIVER CRAWLERS	ISBN 1-882256-43-3
ERIE SHOVEL	ISBN 1-882256-69-7
FARMALL CUB	ISBN 1-882256-71-9
FARMALL F–SERIES	ISBN 1-882256-02-6
FARMALL MODEL H	ISBN 1-882256-03-4
FARMALL MODEL M	ISBN 1-882256-15-8
FARMALL REGULAR	ISBN 1-882256-14-X
FARMALL SUPER SERIES	ISBN 1-882256-49-2
FORDSON 1917-1928	ISBN 1-882256-33-6
HART-PARR	ISBN 1-882256-08-5
HOLT TRACTORS	ISBN 1-882256-10-7
INTERNATIONAL TRACTRACTOR	ISBN 1-882256-48-4
INTERNATIONAL TD CRAWLERS 1933-1962	ISBN 1-882256-72-7
JOHN DEERE MODEL A	ISBN 1-882256-12-3
JOHN DEERE MODEL B	ISBN 1-882256-01-8
JOHN DEERE MODEL D	ISBN 1-882256-00-X
JOHN DEERE 30 SERIES	ISBN 1-882256-13-1
MINNEAPOLIS-MOLINE U-SERIES	ISBN 1-882256-07-7
OLIVER TRACTORS	ISBN 1-882256-09-3
RUSSELL GRADERS	ISBN 1-882256-11-5
TWIN CITY TRACTOR	ISBN 1-882256-06-9

RAILWAYS

CHICAGO, ST. PAUL, MINNEAPOLIS & OMAHA RAILWAY 1880-1940	ISBN 1-882256-67-0
CHICAGO&NORTH WESTERN RAILWAY 1975-1995	ISBN 1-882256-76-X
GREAT NORTHERN RAILWAY 1945-1970	ISBN 1-882256-56-5
MILWAUKEE ROAD 1850-1960	ISBN 1-882256-61-1
SOO LINE 1975-1992	ISBN 1-882256-68-9
WISCONSIN CENTRAL LIMITED 1987-1996	ISBN 1-882256-75-1

TRUCKS

BEVERAGE TRUCKS 1910-1975	ISBN 1-882256-60-3
*BROCKWAY TRUCKS 1948-1961**	ISBN 1-882256-55-7
DODGE TRUCKS 1929-1947	ISBN 1-882256-36-0
DODGE TRUCKS 1948-1960	ISBN 1-882256-37-9
LOGGING TRUCKS 1915-1970	ISBN 1-882256-59-X
*MACK® MODEL AB**	ISBN 1-882256-18-2
*MACK AP SUPER-DUTY TRUCKS 1926-1938**	ISBN 1-882256-54-9
*MACK MODEL B 1953-1966 VOLUME 1**	ISBN 1-882256-19-0
*MACK MODEL B 1953-1966 VOLUME 2**	ISBN 1-882256-34-4
*MACK EB-EC-ED-EE-EF-EG-DE 1936-1951**	ISBN 1-882256-29-8
*MACK EH-EJ-EM-EQ-ER-ES 1936-1950**	ISBN 1-882256-39-5
*MACK FC-FCSW-NW 1936-1947**	ISBN 1-882256-28-X
*MACK FG-FH-FJ-FK-FN-FP-FT-FW 1937-1950**	ISBN 1-882256-35-2
*MACK LF-LH-LJ-LM-LT 1940-1956 **	ISBN 1-882256-38-7
*MACK MODEL B FIRE TRUCKS 1954-1966**	ISBN 1-882256-62-X
*MACK MODEL CF FIRE TRUCKS 1967-1981**	ISBN 1-882256-63-8
STUDEBAKER TRUCKS 1927-1940	ISBN 1-882256-40-9
STUDEBAKER TRUCKS 1941-1964	ISBN 1-882256-41-7

The Iconografix Photo Album Series includes:

CORVETTE PROTOTYPES & SHOW CARS	ISBN 1-882256-77-8
LOLA RACE CARS 1962-1990	ISBN 1-882256-73-5
McLAREN RACE CARS 1965-1996	ISBN 1-882256-74-3

The Iconografix Photo Gallery Series includes:

CATERPILLAR PHOTO GALLERY	ISBN 1-882256-70-0

All Iconografix books are available from direct mail specialty book dealers and bookstores worldwide, or can be ordered from the publisher. For book trade and distribution information or to add your name to our mailing list contact

Iconografix
PO Box 446
Hudson, Wisconsin, 54016

Telephone: (715) 381-9755
(800) 289-3504 (USA)
Fax: (715) 381-9756

CATERPILLAR SIXTY
PHOTO ARCHIVE
Edited by P. A. Letourneau

MORE GREAT BOOKS FROM ICONOGRAFIX

CATERPILLAR SIXTY *Photo Archive*
ISBN 1-882256-05-0

HOLT TRACTORS *Photo Archive*
ISBN 1-882256-610-7

ERIE SHOVEL *Photo Archive*
ISBN 1-882256-69-7

CLETRAC & OLIVER CRAWLERS
Photo Archive ISBN 1-882256-43-3

INTERNATIONAL TRACTRACTOR
Photo Archive ISBN 1-882256-48-4

RUSSELL GRADERS *Photo Archive*
ISBN 1-882256-11-5

LOGGING TRUCKS 1915-1970
Photo Arhive ISBN 1-882256-59-X

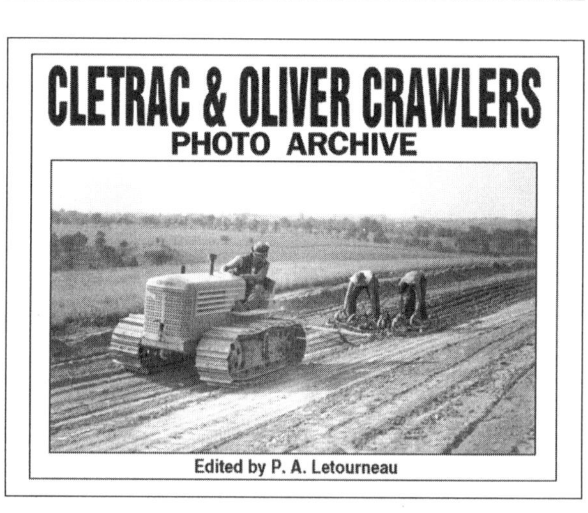